Introduction

To Young Scientists and Their Helpers

Everything that is known about living things, Earth, and the universe has been learned through some type of research or exploration. One way of studying or investigating a problem or question is to perform an **experiment**.

When a question is asked or a problem is posed, an inquiry has begun. A science inquiry can be designed to produce a valid answer to the question asked. Different questions require different designs. In an experiment, a cause-effect relationship is tested. This book addresses the experiment design. Other designs include those for observations, models, collections, and observations.

What is an experiment? An experiment occurs when one variable (the independent variable) is changed. Another variable (the dependent variable) responds to the first and is watched. Other variables (constant variables) remain the same, or are unchanged, throughout the experiment. This probably sounds complicated. However, this book enables an adult and child to have a wonderful time exploring science topics together by providing them with an easy, step-by-step format they can follow to do an experiment.

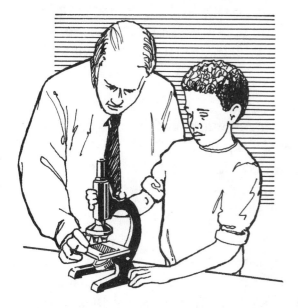

This book helps the young scientist and his or her mentor go through all the steps of conducting an experiment—from choosing what kind of experiment to do to displaying the final product at a science fair. A sample experiment is developed throughout this book to ensure understanding. (The sample experiment involves a child, Connor, who wants to find out the effect of sight on the ability to smell different scents.)

Experimenting is fun. It enables a child to think and use all of his or her senses. You will be delighted as you see a child's self-esteem soar after he or she has reached his or her goal and finished an important task.

1

Planning Calendar and Log Book

Chapter 1

▓ Planning Calendar

Once the child has decided that he or she wants to do an experiment, he or she needs to make a plan. A plan will help the child know what to do and when to do it.

To create a plan, the child first needs to know when the experiment must be completed. Have the child find that date on the calendar and circle it in red. Then, to help the child find out how many days he or she has to complete the experiment, the child can number the dates on the calendar backward. The date circled in red is Day 0. The date before that is Day 1. The child can continue numbering to today's date. This lets the child see how many days he or she has to complete the study.

If the child is doing the experiment for a science fair, most science fairs are held in February through April. This is important in the child's planning. If the child wants to perform an experiment out of doors, he or she needs to do this in favorable weather. Also, remember that some things cannot be studied in particular parts of the country. For example, if the student wants to study how plants survive near the ocean beach, then it is important that you either live near or visit the beach often.

It is also important to know that different experiments take different amounts of time to complete. To help the child decide how long it will take to finish each part of the experiment, have him or her fill out the Planning Calendar on Log It! #1 on page 7. (The Log It! pages have been designed to be completed by the child and can be found throughout the book.) This will be the first page of the child's **Log Book**. (The Log Book is explained in complete detail on pages 3-6.) Remind the child that the filled-in dates are merely suggested times needed for completion.

The Planning Calendar will provide the child with an easy way to see if he or she is on schedule and what tasks still have to be done. When each task is completed, the child should write in that date so that he or she can see what tasks have been accomplished.

Planning Calendar and Log Book

▓ The Log Book

Once the child has set up a tentative plan, the next thing he or she needs to do is set up a Log Book. A Log Book is like a diary or a journal. It is a complete record of the child's work.

The Log Book will contain a variety of information such as a planning calendar, lists of possible topics, information from the library that may be written or photocopied, pictures, information about how to do the experiment, materials needed, and data gathered. Some parts of the Log Book may be printed or written in pencil. Other parts may be written in ink or printed from the computer.

A Log Book should always be present when the child is working on the experiment. By the end of the experiment, it will probably contain water spots or smudge marks. This is accepted and represents the child's hard work.

A Log Book should never be rewritten or recopied. Spelling and grammar does not have to be perfect. The child may use any kind of blank book as a Log Book. He or she may wish to use the Log It! pages found in this book. These pages are found throughout the book, and complete details are given on how to use them. If the child is conducting a science fair study, the Log Book will need to conform with the science fair guidelines. In other studies, the child may decide personally what sections to include.

Scientists have been keeping Log Books for a long time. Theirs, like the child's, are not changed or recopied. Scientists have learned much information from reading earlier scientists' work. We build on information from the past.

Planning Calendar and Log Book

The Log Book serves many purposes. It usually contains the following:

- a daily journal in which to record thoughts, decisions, and reflections
- a time line for the experiment
- background information
- a record of decisions made
- raw data for the experiment
- a day-by-day record of what has been done throughout the experimental study
- a place in which to keep all the important papers about the study

The Log Book should begin the day the study begins. It is important to set up the Log Book in a systematic matter. It should be divided into sections. Use tabs that are purchased or made. The tabs generally used in an experiment study are as follows:

1. Daily Journal
2. Planning Calendar
3. Choosing a Topic
4. Background Information
5. Designing the Experiment
6. Problem and Hypothesis
7. Procedures
8. Data
9. Results and Interpretations
10. Conclusion
11. Sharing the Study

Page 8 contains tabs that the child may wish to use for his or her Log Book. If the child wishes to use these tabs, have him or her cut out each tab. The child can then glue each tab to the right side of a separate blank sheet of paper. The tabs should be positioned so that they are easily visible when in a closed notebook.

Log Book Contents

The Log Book is the story of the experiment. It tells everything that happened from beginning to end. The numbered Log It! pages found throughout this book will assist the child in creating a well documented Log Book. These pages are to be completed by the child and filed in his or her Log Book. In the bottom right corner of each Log It! page is information indicating in which tabbed section of the Log Book the page is to be filed. Regardless of the age of the child, this book will help him or her create a valid piece of scientific research. Below and on page 6 are the recommended sections the child can use in his or her Log Book when doing an experiment. Also included are explanations of each Log It! page that will be a part of these sections.

I. Daily Journal

This area of the Log Book is where the child records his or her day-to-day thoughts, ideas, reflections, and actions.

II. Planning Calendar

Log It! #1 (page 7) is used by the child to determine the plan and time line for the experiment. Categories are listed to assist the child in planning.

III. Choosing a Topic

A child has many decisions to make when choosing an experiment. Chapter 2 deals with how to choose a topic for an experiment. Log It! #2 (page 11) is designed to help the child work through this process.

IV. Background Information

This section helps organize both the search for information and the information collected. Log It! #3 (page 16), *Brainstorming Blast!*, should help the child with his or her search for information about the experiment. There are four types of background information. A Log It! sheet is given for each kind of information. The early researchers or past research dealing with the experiment is included on Log It! #4 (page 17). The importance of the experiment to the child and to mankind go on Log It! #5 (page 18). Facts and terms that explain the topic go on Log It! #6 (page 19). The procedures, or ways to do the experiment should be placed on Log It! #7 (page 20). Log It! #8 (page 21) provides the child with a place where he or she can analyze each piece of background information that was gathered. Make as many copies of this page for the child as needed.

Log Book Contents continued

V. Designing the Experiment

This section guides the child through the process of designing the experiment. Log It! #9 (page 27) is where the child records this design.

VI. Problem and Hypothesis

The problem in an experiment is the important question upon which the experiment is based. This should be recorded on Log It! #10 (page 28) along with the hypothesis. The hypothesis is an educated guess as to how the question or problem will be answered.

VII. Procedures

It is in this section that the child will describe the steps and the set-up of the equipment and/or materials for the experiment. In addition, he or she will describe the variables. The variables should be recorded on Log It! #11 (page 29). The setup and materials needed are facilitated through Log It! #12 (page 33).

VIII. Data

The child needs to record data as he or she collects it while doing the experiment. This data should be recorded in the sample data table on Log It! #13 (page 34).

IX. Results and Interpretations

The child will need to prepare a graph based on the data he or she collected. This should be recorded in the graph found on Log It! #14 (page 38). The interpretation of the graphed data should be recorded on Log It! #15 (page 39).

X. Conclusion

Log It! #16 (page 42) provides an area in which the child can write a clear, concise conclusion.

XI. Sharing the Study

Display Design Decision, Log It! #17 (page 47), aids the child in planning a visual display for a science fair or classroom presentation.

◢ MY PLANNING CALENDAR   Date _____

▨ Essential Information

Rules and Regulations Obtained	YES	NO
Assessment Criteria Obtained	YES	NO
Local Fair or Assignment Due Date		_____
District Fair Due Date		_____
Regional Fair Due Date		_____

▨ Success Calendar

	Planned Date	Date Completed
My Countdown Calendar (1 day)	_____	_____
Setting Up My Log Book (1 day)	_____	_____
Choosing a Topic (2-5 days)	_____	_____
Collecting Background Information (1-3 weeks)	_____	_____
Problem and Hypothesis (1-4 days)	_____	_____
Set-Up or Design for Experiment (1 week)	_____	_____
Getting Materials Ready for the Experiment (1 week)	_____	_____
Making the Data Table (1-2 weeks)	_____	_____
Recording in the Data Table (1-2 weeks)	_____	_____
Stating Results (1 week)	_____	_____
Drawing Conclusions (1 week)	_____	_____
Compiling a Bibliography (2-3 days)	_____	_____
Making the Display (1-2 weeks)	_____	_____

Log It! #1
Section: Planning Calendar

Daily Journal

Procedures

Planning Calendar

Data

Choosing a Topic

Results and Interpretations

Background Information

Conclusion

Designing the Experiment

Sharing the Study

Problem and Hypothesis

Choosing a Topic

When trying to decide on a topic for the experiment, first have the child pick a topic or category that he or she wants to learn more about. If ideas are needed, the child can think of things that he or she has done at school or home. For example, perhaps the child saw butterflies in the yard and wants to know more about them. You can have the child look for ideas in the library or talk with a neighbor. Some examples of possible experiment topics are listed below. Also listed are possible problem questions related to each topic that the child would try to solve.

Science Category	Possible Topic	Problem
Behavioral Science	Memory	How good is your visual memory?
	Food Birds Prefer	Do birds like some foods better than others?
Botany	Seed Growth	Do some seeds grow faster than others?
	Water and Plants	How much water do plants need to grow?
Chemistry	Fruit Spoiling	Can you keep apples from turning brown?
	Substances and Water	What will dissolve in water?
Earth and Space Science	Shadows	Does your shadow change during the day?
	Freezing Water	At what temperature does water freeze?
Environmental Science	Soil Changes	What happens when soil washes downhill?
	Communities	What lives in a soil community?
Medicine and Health	Hearing	Can you identify an object by sound only?
	Smell	Can you smell what you cannot see?
Microbiology	Mold	How does mold grow on bread and fruit?
	Pond Life	What lives in a pond?
Physics	Movement	What makes things move?
	Sounds	What sounds are made when things move?

Choosing a Topic continued

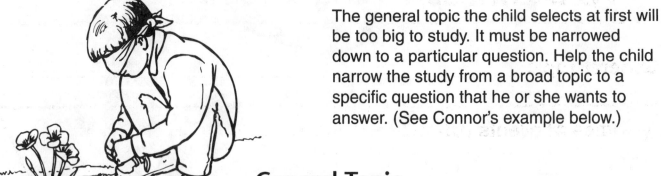

The general topic the child selects at first will be too big to study. It must be narrowed down to a particular question. Help the child narrow the study from a broad topic to a specific question that he or she wants to answer. (See Connor's example below.)

General Topic
MEDICINE AND HEALTH

(Child should choose only one.)

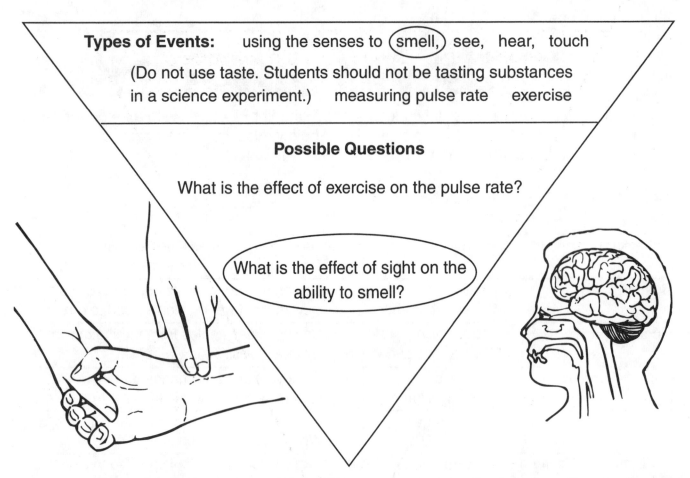

Types of Events: using the senses to (smell,) see, hear, touch

(Do not use taste. Students should not be tasting substances in a science experiment.) measuring pulse rate exercise

Possible Questions

What is the effect of exercise on the pulse rate?

What is the effect of sight on the ability to smell?

In the sample experiment, Connor chose Medicine and Health as the general topic. Connor wants to use the sense of smell, and the possible question is "What is the effect of sight on the ability to smell?"

Log It! #2 (page 11) has been designed to help the child select and narrow down his or her topic. Have the child fill it out and put it in the Log Book.

GETTING DOWN TO BUSINESS

Date _____

General Topic: _____

Types of events (Choose and circle one.):

Problems to Study (Possible Questions):

Log It! #2
Section: Choosing a Topic

FS-62105 Experiments

Gathering Background Information

Once the child has chosen a topic, he or she will need to find a lot of information about that topic. This is called collecting **background information**. Background information will assist the child in developing a problem or question for his or her experiment. This information may be gathered from a variety of sources. There are four different kinds of background information. They are listed below and on page 13. Be sure to discuss each of these with the child.

▓ Earlier Scientists

This is what is known about the topic from the past. Did any scientists study the topic in the past? Is there a famous study about the topic? *(Example: Connor would look to see what is known about smell and sight.)*

▓ Importance

This background information suggests what is important about the study. Why is studying this topic important to the child and to the world? Could it help the environment? Could it help others understand how machines work? *(Example: Connor would find out how the study is important to the understanding of smell and sight.)*

▓ Factual Information

This is true information about the topic. It might tell the child of what something is made. Facts might tell where something is found. If the child needs to find information about a living thing, facts would tell him or her where it lives and what it eats.
(Example: In the experiment that Connor is doing: "Can you smell pepper, garlic powder, vanilla, and cinnamon when you cannot see them?", Connor would find factual information on the nose and smelling, the eye and sight. Connor would also find information on the different scents.)

▓ Procedural Information

Procedural information tells what to do and how to do it. The child can learn what materials are needed for the experiment. He or she can learn how to do the experiment. The child may answer questions like these: How long will the materials last? Are the materials safe to use? What is the best way to work with the materials? *(Example: In Connor's experiment, Connor would find out how to work with the different scents. Connor would make certain that the scents chosen are safe to use.)*

This may seem like a lot of information to include. However, it is important in order to help the child understand how to do an experiment. A young child may want to start "doing the experiment." That is okay. The child will begin asking questions. Then he or she will begin searching for the facts and other background information. The child will probably gather background information throughout the study as questions arise.

The Background Information section will look different from the other sections of the child's Log Book. It can include all or part of the following:

- photocopies of information that the child found,
- tape recordings of people the child interviewed,
- drawings made by the child,
- pamphlets received,
- notes the child wrote,
- pages printed from a CD disk on a computer, or
- pages printed from the Internet.

There may be hole-punched pages, pages with glued articles, yellow highlighted sentences, scribbled notes, sketches or drawings, splotches, splashes, and smudges! Each of these types of information is important and add depth and character to the child's study.

Getting Started

So where and how does the child start gathering background information? A fun and easy way to begin is by brainstorming. This is a great way for the child to find out what he or she already knows about the topic for the experiment. Encourage the child to ask friends, parents, teachers, and others to help with the brainstorming. The goal should be to help the child find out as much about the experiment as possible. Give the child Log It! #3 (page 16) and have him or her write down all the words that he or she and others know about the chosen topic. Be sure the child writes down words which describe or are related to the topic.

When the child has finished writing down every word that he or she, or others can think of, it is time to fit them into one of the four kinds of background information or to cast them aside.

When categorizing, the first step is to have the child code the words that he or she wrote on Log It! #3. To do this, have the child scratch out words that do not fit and color code the important words using crayons or markers. Words can be color coded as described below.

- Words relating to earlier scientists can be circled in blue and recorded on Log It! #4 (page 17).
- Words relating to importance can be circled in black and recorded on Log It! #5 (page 18).
- Words which are facts can be circled in green and recorded on Log It! #6 (page 19).
- Words which relate to procedures can be circled in red and recorded on Log It! #7 (page 20).

The child should consider the places below when gathering information.

People	Places	Things
family/friends	home	magazines, books, telephone white pages
professionals	government	telephone blue pages
	businesses	telephone yellow pages
	associations	telephone yellow pages
teachers	school	textbooks
librarians	library	books, tapes, Internet
computer experts	computer lab	computer CDs
sales persons	bookstores	books, computer CDs
scientists	museums	apparatus, books, pamphlets
	research labs	books, apparatus
	universities	interviews

Recording Background Information

The child has now developed good plans for collecting background information with the help of Log It! pages #4-#7. Reading and recording the useful information is the next logical step. This section becomes complete when the child has answered the questions located at the top of these four Log It! pages.

Good information is sometimes difficult to identify. A good indicator of useful information is if two different sources repeat such things as a person's name and specific characteristics. Suggest that the child photocopy useful information. The child can then use it when it is most useful. He or she may wish to highlight what is most useful and glue it to a copy of Log It! #8 (page 21). (Make as many copies of this page as needed for the child. If a photocopier is not available, help the child record only the most useful information.)

▧ Adding Reference Data

In addition to making notes of the materials read, have the child include the reference data. Log It! pages #4-#7 have a bibliography, or reference section, the child can complete to show where the information was found. The chart below shows examples of needed reference data and how to set it up.

TYPE	AUTHOR/PERSON	SOURCE
Interview	*Name*	*Position, address, phone*
Example	Dr. Kevin Jones	Physician, 94 Parkway St., 555-8506
Book	*Author*	*Copyright date, Book Title*
Example	S. Smith	1994, <u>Smell</u>
Magazine	*Author, Article title*	*Magazine date, Magazine*
Example	Sam Smith, "How We Smell"	May, 1993, <u>Science Facts</u>
Encyclopedia	*Term or topic*	*Encyclopedia Name, Copyright*
Example	Smell	<u>World Book</u>, 1990

Have the child make a comprehensive list of reference data using all of the references found on each Log It! #4-#7 page. The data should be categorized by type and should be in alphabetical order by person. This reference sheet should be placed at the back of the Background Information section in the child's Log Book.

Write down words that describe or that are related to the experiment you will be doing. Ask parents, friends, teachers, and others to add words to the brainstorming circle. Don't worry about spelling. Write quickly and scatter the words about the circle.

The topic that I am studying is _____ .

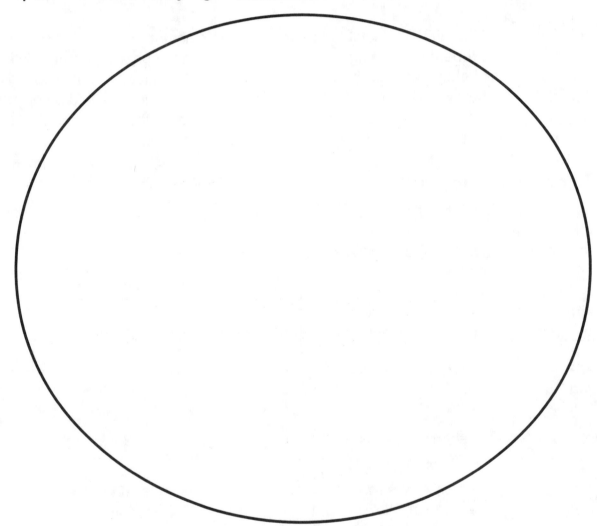

- Circle all the words relating to Historical Information in blue.
- Circle all the words relating to Importance Information in black.
- Circle all the words relating to Factual Information in green.
- Circle all the words relating to Procedural Information in red.

Any words that do not seem to relate can be crossed out.

Log It! #3
Section:
Background
Information

BACKGROUND INFORMATION Date _____
Earlier Scientists and Other History

This page is for information about earlier scientists and/or early scientific information. Record below the blue circled words from page 16. Define these terms if necessary. Answer the following questions: Did a famous scientist work with this topic? What is known about those studies?

Past Scientists and/or Past Information:

_____

_____

_____

_____

_____

_____

_____

_____

_____

_____

_____

_____

Possible Sources of Information:

Type (Book, magazine, encyclopedia, Internet, interview, etc.)	Author/Person	Source (Include page numbers or phone number.)

Log It! #4
SECTION: Background Information

BACKGROUND INFORMATION Date _____

Importance to Me and to Mankind

This page is for recording the importances of the experiment to you, others, and other things. Record the black circled words from page 16. Define any necessary terms. Answer these questions: How is the study important to you? Is the study important to other people? Is it important to the health of others? Is it important to the environment?

Terms   **Importance**

_____

_____

_____

_____

_____

_____

_____

_____

_____

_____

_____

Possible Sources of Information:

Type (Book, magazine, encyclopedia, Internet, interview, etc.)	Author/Person	Source (Include page numbers or phone number.)

Log It! #5
Section: Background Information

BACKGROUND INFORMATION Date _____

Facts About My Topic

This page is used for recording facts. Record the green circled words from page 16. Define any necessary terms. Answer these questions: What are the most important terms or ideas about the topic? How would you explain or describe them?

Terms   **Description**

_____

_____

_____

_____

_____

_____

_____

_____

_____

_____

_____

Possible Sources of Information:

Type (Book, magazine, encyclopedia, Internet, interview, etc.)	Author/Person	Source (Include page numbers or phone number.)

Log It! #6
Section: Background Information

BACKGROUND INFORMATION Date _____
Procedures

This information relates to how the experiment can be done and what materials or supplies are needed. Record the red circled words from page 16. Define any necessary terms.

Ways To Do Experiment

_____
_____
_____
_____
_____
_____
_____
_____
_____
_____
_____

Supplies Needed

_____
_____
_____
_____
_____
_____
_____
_____
_____
_____
_____

Possible Sources of Information:

Type (Book, magazine, encyclopedia, Internet, interview, etc.)	Author/Person	Source (Include page numbers or phone number.)

Log It! #7
Section: Background Information

NOTES PAGE

Date _____

Reference Data		
Type	Person	Source

Type of Background Information
Circle One

History   Importance   Facts   Procedures

Recorded Notes

Log It! #8
Section: Background Information

Designing the Experiment

Once the child has gathered enough background information, he or she is ready to design the experiment. This is the point at which the child describes exactly what he or she wants to find out and how he or she will find out. The child can use the background information collected as a guide. Then the child needs to develop an experimental design which includes the following three components: **Problem**, **Hypothesis**, and **Procedure**. It is important that these three parts are accurate and that they are kept simple.

To help the child keep the experiment simple, the child should study one thing or problem at a time. This is very important, especially if the child is doing an experiment for the first time.

Help the child think of designing the experiment as writing a story. A story has these parts or elements:

> **main characters** — what or who the story is about
>
> **setting** — where, when, and how the story takes place
>
> **plot** — what happens

Scientists do not use these terms, but they do use these ideas. The scientist's terms are found on Log It! #9 (page 27). To complete this page, the child must look at the facts and procedures portion of the background information. He or she must organize this information into the proper story elements. At this point, the child is ready to select the main character, setting, and plot. However, go over the example provided on page 23 with the child before he or she completes Log It! #9 (page 27).

Design Example

Connor decided to try to learn if a person can smell certain scents when blindfolded. Connor will use the encyclopedia and CD-Rom information on "smell" to help with the design of the experiment. Connor has decided to use scents that are easily identified and found around the home.

Design Decisions
Setting (Constant Variables)

WHERE—The experiment could be conducted in my kitchen at home or in the classroom.	WHEN—The experiment could be conducted at any time of the year.	HOW—I will obtain five cotton balls. These will be dipped into the scents and then placed in plastic bags. I will get five plastic bags. There will be one for each cotton ball. I will need pieces of cloth to use as blindfolds. These items—cotton balls, plastic bags, and blindfolds—will be used throughout the experiment.

Main Characters
(Independent Variables)

Plot
(Dependent Variables)

Pepper	Record whether the student can or cannot smell the scent.
Garlic powder	Can smell the scent / Cannot smell the scent
Vanilla	Can smell the scent / Cannot smell the scent
Cinnamon	Can smell the scent / Cannot smell the scent
Plain piece of cotton	Can smell the scent / Cannot smell the scent

Specifying the Problem

Completing the design is a major task. Congratulate the child for doing a good job! Once this has been done, the child is now ready to move on to the **problem**, **hypothesis**, and **procedure**. The child will need to use Log It! #10 (page 28) to help do this.

The child should begin by writing the **problem**. The problem is a question that asks what is to be accomplished in the experiment. The question will do the following:

- state the purpose of the experiment,
- list the items to be studied (independent variables),
- tell how to study (observe or measure) the items (dependent variables),
- explain under what conditions the items will be studied (constant variables).

In Connor's experiment, the problem would be written as follows:

> *Can you smell or not smell pepper, garlic powder, vanilla, cinnamon, or a plain cotton ball when you are blindfolded?*

Connor's independent variables are the pepper, garlic powder, vanilla, cinnamon, and plain cotton ball. The plain cotton ball represents the **control** of the experiment. The child always needs a control. This is a baseline point. It is something that does not change. It allows the child to compare the changes in other things to it.

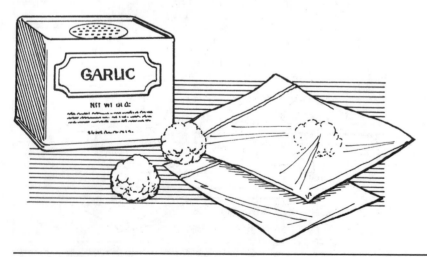

The dependent variables in Connor's experiment are **to smell or not smell the scents**.

The constant variable is the blindfold. The cotton balls and plastic bags are also constant variables.

Have the child write his or her problem on Log It! #10 (page 28).

Writing the Hypothesis

The child is now ready to write the **hypothesis**. The hypothesis is an educated guess as to what the child thinks the experimental results will be. What results will the child get? This should not be a wild guess. The hypothesis should be based on the background information.

Connor decided that the hypothesis should be as follows:

> **I guess** that you can smell pepper, garlic powder, vanilla, and cinnamon even when you are blindfolded and cannot see them **because** you can smell what you cannot see.

The format is always an **I guess . . . because** statement. The **I guess** portion lists the main characters, or the independent variables, of the experiment. The **because** part of the hypothesis lists the dependent variables, or plot, of the story. This predicts what is going to happen as a result of the independent variables. The constant variables, or setting, should also be included in the hypothesis.

At this time, have the child write his or her hypothesis on Log It! #10 (page 28).

Setting Up Procedures

The child is now ready to make another big decision. He or she must plan how to do the experiment. These are the directions. When a new bike is purchased, it has to be put together. Directions tell you how to do that. The directions need to be simple. Anyone should be able to follow the child's directions. The directions or procedure can be presented in written form or with drawings and labels. It may also be a combination of drawings and written directions.

The procedure should have step-by-step directions for performing the experiment. Have the child look at Connor's sample below.

It is helpful to use a sketch and a set of directions when determining procedures. A picture can be substituted for a sketch. The directions may or may not be numbered. Each type of procedure could be recorded on the same type of sheet.

SKETCH

DIRECTIONS

Connor's procedure would look like this: Put some pepper on one cotton ball. Place it in the plastic bag. Do these same steps with the garlic powder, vanilla, and cinnamon. Be sure to put a plain cotton ball in a plastic bag to serve as the control. Blindfold a student. Hold the bag containing the pepper near a student's nose. Move the open bag back and forth. Ask the student to tell you what he or she smells. Record his or her answer in the data table. Repeat these steps with each scent. Then repeat the entire process with the next student until all students have had a chance to do the experiment.

Have the child complete Log It! #11 (page 29) at this time.

READY, SET, GO!
DESIGN DECISIONS

Date _____

Setting (Constant Variables)

WHERE	WHEN	HOW

Main Characters
(Independent Variables)

Plot
(Dependent Variables)

Log It! #9
Section: Designing the Experiment

Q & A! MY PROBLEM AND HYPOTHESIS

Date _____

The problem for my experiment is as follows:

_____

_____

_____

_____

_____

_____

The hypothesis being tested by my experiment is as follows:

_____

_____

_____

_____

_____

_____

_____

_____

_____

Log It! #10

Section: Problem and Hypothesis

SETTING UP
MY PROCEDURES

The type of procedure being designed is a

_____.

SKETCH	DIRECTIONS

Log It! #11
Section: Procedures

Carrying Out the Experiment

Chapter 5

The child has finished designing the experiment. Now he or she is about ready to perform the experiment. This is what is meant by "carrying out the experiment." Before beginning the experiment, a few decisions still have to be made. They are listed below.

■ Decision 1—Samples

The independent variables, or main characters, are the child's samples. How many samples of each one does he or she need? Suppose that the child is growing plants. The child would need at least five samples of each independent variable. More would be good. If the child had only one or two plants, something might happen to them. Then he or she would not have any results. *(Connor needs only one sample of each independent variable. In other words, Connor needs only one sample of pepper, garlic powder, vanilla, cinnamon, and plain cotton.)*

■ Decision 2—Where to Do the Experiment

Where is the best place for the child to perform the experiment? Is it best to do it outside, in the kitchen, or in the basement? Maybe it could be done in the classroom. A scientist might even be willing to share his or her laboratory. *(Connor would do the experiment in the classroom.)*

■ Decision 3—Trials

The number of trials refers to how many times the child will repeat the experiment. The results will be better if the child does the experiment several times. *(Connor might do the experiment in the classroom and have each student try to smell the scents.)*

■ Decision 4—Materials Needed

The child now needs to gather any necessary materials. Does the child have them at home? Does he or she need to obtain materials from the grocery store? Have the child make a list of materials needed. If the child needs permission to use something, be certain that he or she obtains it now.

The child should now complete Log It! #12 (page 33), *Decisions, Decisions*.

Making a Data Table

For any experiment, the child needs to construct a **data table**. A data table is a place where scientists place the information that they gather from the experiment. This information is called **data**.

Explain to the child that the data table organizes the information that has been gathered from doing the experiment.

Tables are made up of horizontal and vertical lines that are organized into a grid. Explain to the child that horizontal lines go from the left to the right side of the page. The space between them is called a **row**. Vertical lines go from the top of the page to the bottom of the page. The space between vertical lines is called a **column**. Show the child the graphic below.

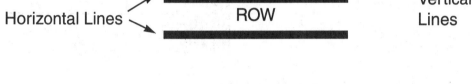

Horizontal Lines

ROW

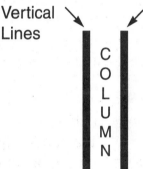

Vertical Lines

COLUMN

Below are some terms to help the child become familiar with data table components.

- The title of any table is the plot (dependent variable).
- The first row is named for the general topic.
- The next row is where the columns are named. These are the main characters (independent variables).
- The first column should contain the number of trials. It might be numbered 1-10 or lettered A-J.
- The remaining columns are where the child records the information or data.

Have the child look at the sample data table on page 32. Then the child should record his or her data on Log It! #13 (page 34).

Making a Data Table continued

Sample: Connor's Data Table
Can or Cannot Smell the Scents

		Scents (General Independent Variable)				
		Pepper	Garlic Powder	Vanilla	Cinnamon	Control (Plain Cotton)
	1	Can Smell	Cannot Smell	Can Smell	Cannot Smell	Cannot Smell
	2	Can Smell	Can Smell	Can Smell	Cannot Smell	Cannot Smell
S	3	Cannot Smell	Can Smell	Cannot Smell	Can Smell	Cannot Smell
T	4	Can Smell	Can Smell	Cannot Smell	Can Smell	Cannot Smell
U	5	Cannot Smell	Can Smell	Can Smell	Cannot Smell	Cannot Smell
D	6	Can Smell	Can Smell	Can Smell	Can Smell	Cannot Smell
E	7	Can Smell	Cannot Smell	Cannot Smell	Can Smell	Cannot Smell
N	8	Can Smell	Can Smell	Can Smell	Cannot Smell	Cannot Smell
T	9	Can Smell	Can Smell	Can Smell	Cannot Smell	Cannot Smell
S	10	Can Smell	Can Smell	Cannot Smell	Can Smell	Cannot Smell
	11	Cannot Smell	Can Smell	Can Smell	Cannot Smell	Cannot Smell
	12	Can Smell	Cannot Smell	Can Smell	Can Smell	Cannot Smell
	13	Can Smell	Can Smell	Cannot Smell	Can Smell	Cannot Smell
	14	Can Smell	Can Smell	Can Smell	Can Smell	Cannot Smell
	15	Can Smell	Can Smell	Can Smell	Cannot Smell	Cannot Smell

At the end of the data table, the child may need to **total** the columns. Explain to the child that this means to add all the numbers in that column. You may need to help the child accomplish this task. If the child needs to average the numbers and has not yet learned how to do this, have him or her put all the numbers in a column from lowest to highest.

1
2
3
4
5
6
7
8
9
10

Then have the child fold the paper at the middle number. This is the average.

◾ DECISIONS, DECISIONS

Date _____

Number of samples:_____Why?_____

_____

_____

_____

_____

_____

Where I am going to do the experiment: _____

_____

_____

_____

_____

_____

Number of trials:_____

Materials needed: Where obtained:

_____

_____

_____

_____

_____

_____

Log It! #12
Section: Procedures

Log It! #13
Section: Data

Studying the Results

The child is almost finished with the experiment. He or she has finished collecting data. All of the data should be in the data table. Now the child is ready to study the data and see what it all means.

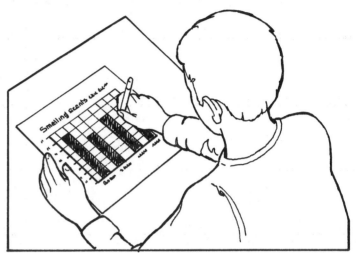

■ Making a Graph

The first part of studying the results is to make a graph. A **graph** is a picture of the results. It is often easier to understand pictures than words or numbers.

The child may wish to make a graph in one of the following three ways:

1. Buy graph paper. It already has the grid lines.
2. Make his or her own graph using a ruler to make the lines. This is difficult because it is hard to make straight lines.
3. Have someone show him or her how to make a computer graph or chart.

There are some basic rules that need to be followed when making a graph. They are as follows:

- If the child's items to be studied are in word form (independent variable), make a bar graph.
- If the child's items to be studied are in number form (independent variable), make a line graph.
- The horizontal axis is the line going across, or from left to right, on the page. The horizontal axis is where the independent variable should be placed.
- The vertical axis is the line going from the top to the bottom of the page. The vertical axis is where the dependent variable should be placed.

The graph should have a title. The title should connect the independent and dependent variables. *(In Connor's experiment, the title would be "The effect of certain scents on the ability to smell what you cannot see." This can be simplified by saying "Smelling scents when you cannot see them.")*

Studying the Results continued

The child should begin numbering with 0 at the lower left corner of the graph. If the child determined averages on the data table, this is what he or she should enter on the graph.

Below is a diagram of how the graph should be set up:

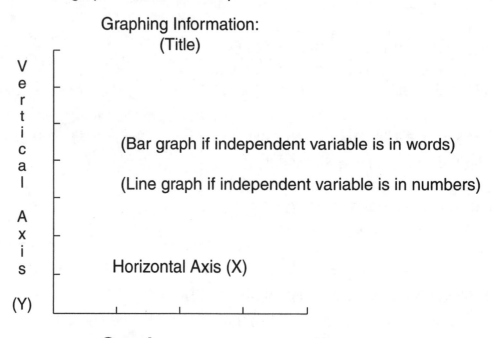

Graphing Information:
(Title)

V
e
r
t
i
c
a
l

A
x
i
s

(Bar graph if independent variable is in words)

(Line graph if independent variable is in numbers)

Horizontal Axis (X)

(Y)

Graph

Students Smelling/Not Smelling Scents They Cannot See

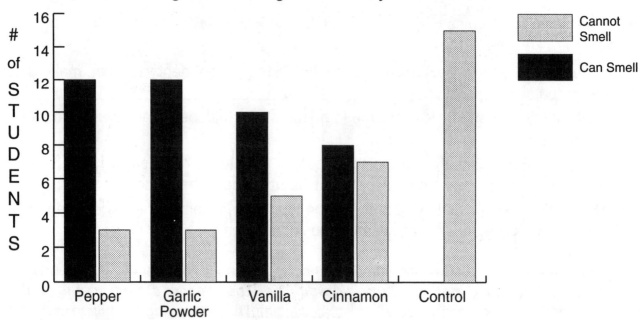

Have the child draw an appropriate graph on Log It! #14 (page 38).

Interpretation

The next step for the child concerning the results section of the Log Book is to interpret or figure out what the graph means. What do those bars or lines on the graph really represent?

Below are some questions that may help the child figure it all out.

- Which independent variable(s), or main character(s), shows the most of what you are studying? Which has the highest bar or line? *(In Connor's experiment, the garlic and pepper are tied. Twelve people could smell each.)*

- Which independent variable(s), or main character(s), shows the least of what you are studying? Which has the lowest line or bar? *(In Connor's experiment, the pepper and garlic were tied. Three people could not smell each.)*

- Which independent variables(s), or main character(s), was in between the highest and the lowest line or bar? *(In Connor's experiment, ten people could smell the vanilla while only eight could smell the cinnamon. Seven people could not smell the cinnamon while only five could not smell the vanilla.)*

Give the child Log It! #15 (page 39) to use to write an interpretation of his or her experiment results.

◼ MY GRAPH

Date _____

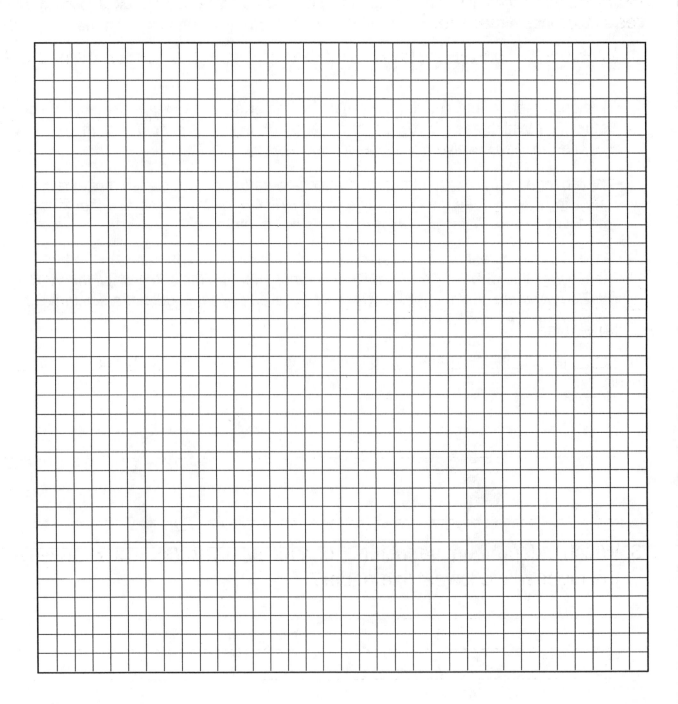

Log It! #14
Section: Results and Interpretations

▪ INTERPRETATION

Date _____

Answer the questions below to help you understand your graph and the results of the experiment.

1. Which independent variable(s), or main character(s), has the highest line or bar?

_____

_____

_____

_____

_____

2. Which independent variable(s), or main character(s), has the lowest line or bar?

_____

_____

_____

_____

_____

3. Which independent variable(s), or main character(s), was in between the highest and lowest line or bar?

_____

_____

_____

_____

_____

Log It! #15
Section: Results and Interpretations

Writing the Conclusion

Chapter 7

The **conclusion** is the last part of the experiment that the child must write. It is one of the most important parts. This is where the child explains his or her results. The conclusion involves relating the experiment to the entire scientific process that has been developed throughout the course of this book.

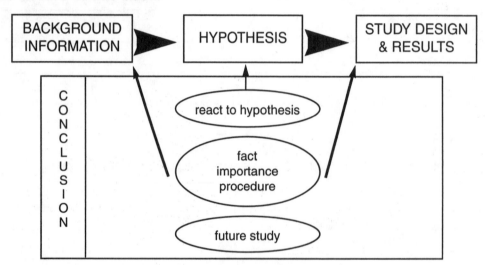

There are five short parts to the conclusion. These include the following:

1. a reaction to the hypothesis and the reason for the reaction,
2. a comparison of the child's study to the background facts,
3. the importance of the experiment to mankind,
4. checking the procedures—what worked and what did not,
5. a prediction of any similar future studies that are planned.

These parts are further described below and on page 41.

■ 1. Reaction to the Hypothesis

There are three possible reactions. To determine the correct answer, have the child ask himself or herself the following question: "When I compare my study findings to the hypothesis, I see . . ."

- **support**—The characteristics which I chose to use worked to solve my problem or question. (The child should provide an example or two which supports this choice.)

- **lack of support**—The characteristics which I chose to use did not work well to solve my problem or question. (The child should provide an example or two which supports this choice.)

- **uncertain support**—Some of the characteristics which I chose to use were helpful and some were not. (The child should provide an example or two which supports this choice.)

40 FS-62105 Experiments

(In Connor's experiment, the results match, or support, the hypothesis. Connor thought that people could smell scents that they could not see. This turned out to be true. In all cases, more people could smell the scents than could not.)

■ 2. Comparison of the Child's Study to the Background Facts

The child now should revisit the background information. Did his or her experimental results agree with what is known about the subject? Was anything new or different from the expected results found during the study? Answers to these questions relate the experiment to the factual basis of the original study design.

■ 3. Importance of the Experiment to Mankind

Have the child once again look at the background information. He or she should look at the importance of the experiment. Are the experimental results important to the child? Are they important for one's health or for such things as the environment?

■ 4. Checking the Procedures

The child should look at the experimental design for strengths and weaknesses. The Log Book may have comments like, "I wish I had done this, or I wish that I had chosen that . . ." The strengths and weaknesses should be listed with reasons for these comments. You and the child may wish to look at page 48, *Assessing the Study* in the Appendix. This page can serve as a self evaluation tool for the child as well as an evaluation tool for a mentor or teacher.

■ 5. Prediction of Future Studies

If the child is considering continuing the study, he or she should comment on what would be studied. He or she might like to pursue another aspect of the topic or improve the design.

Give the child Log It! #16 (page 42) so that he or she can write a conclusion containing all five parts.

◪ MY CONCLUSION   Date _____

1. The hypothesis of my study was _____
   _____

   Based upon my study,   **I'm uncertain about**   **I lack**   **I have**
   support for my hypothesis because _____
   _____
   _____

2. When I compare my results to facts in background information,
   I find _____
   _____
   _____

3. I believe my experiment is important because _____
   _____
   _____

4. In assessing my study design, I have found these strengths: _____
   _____
   _____

   and these weaknesses: _____
   _____
   _____

5. I am   **planning**   **not planning**   to continue my experiment
   because _____
   _____
   _____

Log It! #16

Section: Conclusion

FS-62105 Experiments

Sharing the Study

Chapter 8

Sharing an experiment can be a lot of fun. The child's experiment will be interesting for others to see. The child now knows more about this topic than his or her classmates. The Log Book has lots of valuable information that the child will want to share with others. Informally, the child can share with friends what he or she has done. Formally, the child can share the study in one of three ways. In each of these ways, be certain that the child includes the background information, the problem, the hypothesis, the design procedures, the data, the results, and the conclusion. The three ways to share the study are as follows:

- **Preparing a visual display**—This is a clear and visually interesting report. Viewers can easily see what the child did and what he or she found out. This is usually required for science fairs.

- **Writing a report**—This is an exacting report written for other scientists. By reading it, others could follow the procedure and repeat the study. The report could also be written for a class.

- **Giving a talk**—This focuses on what is known, what the child did, and what the child found out. A child could give a talk in class.

A title will be needed for the study. Help the child select a short, interesting one. Connor's experiment could have titles such as *A Smelly Experiment, What's That Smell?, A Nose Knows,* etc.

▓ Visual Display

When preparing a visual display, the child needs to realize that science fairs differ in the amount of space that can be used. Most fairs allow a space of at least 60 cm width by 60 cm depth by 120 cm height. This is almost 2 feet by 2 feet by 4 feet. Have the child make his or her display slightly smaller than the display area permitted just to be safe.

Sharing the Study continued

Below are things to consider when making a display:

- Lettering should be easily read from a distance.
- Materials should be arranged to tell a story like a book, left to right, top to bottom.
- The independent variables (main characters) and dependent variables (plot) should be included in each section. The constant variables (setting) are always located in the procedures section and may be included in other sections. The scientific variable terms should be included in the reporting process.

- Color should be used to attract attention. Having each characteristic a different color aids comprehension.
- Arrangement of information should be interesting and space should be used carefully.
- The display should be lightweight and be able to be folded flat.
- The display should be able to be quickly and easily assembled.
- The display should have an attached base.

Most displays have three sides, a base, and a title board. Science studies are frequently arranged as seen below. The child can make a 1-, 2-, or 3-sided display. He or she should be certain that it fits within the space allowed. Be sure that the sides are spread open enough so that it can be easily read from the front. Use either a large middle board with smaller side boards or a small middle board with larger side boards.

Title		
Background Information	Procedures	Results
	Data	
Problem Hypothesis		Conclusion

Report
and
Log Book

While the child is planning the visual display, have him or her look at advertisements in newspapers and magazines. This will give him or her ideas on how to use space, color, and letter size. Another place to look is in science centers or zoos. They sometimes have experiments on display. Photographs of science displays are another possible source of information.

Something else the child should check on is whether or not computer printing can be used. It is often much easier for the child to enter experimental information into a computer. Help the child choose a font that can be read easily from a distance. A large font size should be used, and the style should be bold. Have the child use spell check, and help him or her correct any errors and reprint before the information is glued to a colored backing. If a computer cannot be used, the child can use a bold, felt-tipped pen for printing. Overhead transparency pens work well. They come in different colors, print boldly, and are easily held. Have the child print on different sizes of unlined sheets, NOT directly on the posterboard. Lined sheets can be used if needed in the primary grades. Give the child different sized sheets to write on.

Teacher, office, and art supply stores are great places where the child can find needed materials for the display. Purchase an inexpensive display board. The child will have choices of color. Have the child use construction paper or thin posterboard for colored backing of the printed material. A good choice for glue is a glue stick.

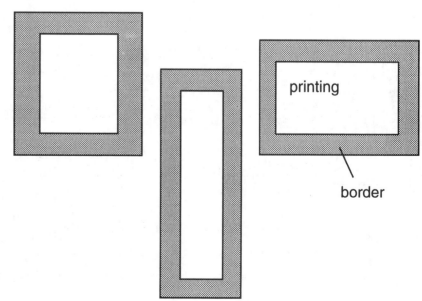

printing

border

Give the child a copy of Log It! #17 (page 47) to help him or her create a visual display. Be sure that the child has looked at enough advertisements and displays to know how to make good decisions.

■ Report

A report includes all the information the child has on his or her display. However, in a report, this information is shared in greater detail. The experiment is shown via photographs in the report. The report is explicit enough so that another person could repeat the same study and check his or her findings with the child's findings. Be sure the child includes the characteristics used and the different appearance of each of them.

■ Talk

A talk includes most of the information that the child has on the display. This is a "show and tell" type of reporting. Tell the child to watch a TV news report of some event to get ideas for his or her talk. A report of this nature usually begins with "how I became interested in the experiment." Drawings of the graphs can be used to show how the experiment was performed. Have the child describe the results. The child can explain what the results mean. The child can end the talk by encouraging others to become interested in experimenting and explain how the experiment might be improved in a future study.

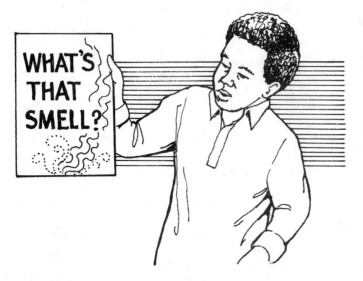

The experiment study has come to an end. Hopefully, the child will have a great feeling of accomplishment. Congratulate him or her for a job well done! The child has worked as a scientist. This may be the first of many experiments that he or she will perform. The child has learned how to organize his or her time and how to complete a task. By working with you, the child has developed skills that are valuable when working with other people. These skills will help the child in all that he or she does.

DISPLAY DESIGN DECISION

Date _____

Below are some decisions you must make when creating a visual display for your experiment study.

The title of my display is _____

Sections which will be on my boards are:

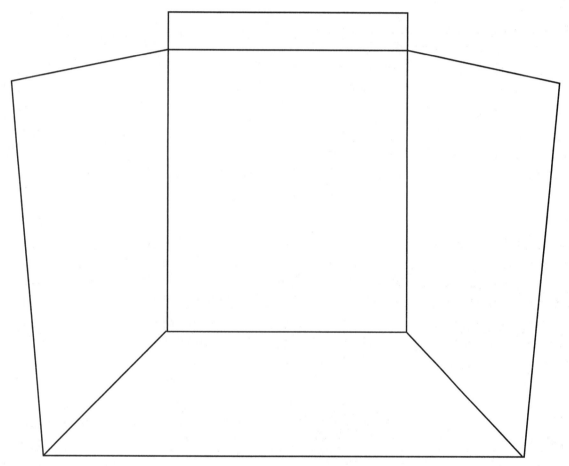

How will the experiment be displayed? _____

_____

_____

Color of display board: _____

Color of backing for white, printed sheets: _____

Color of main characters: _____

Log It! #17
Section: Sharing the Study

Assessing the Study

DISPLAY

Points	Criterion	Explanation	Rating
0 - 2	Easily Viewed	Display faces forward, materials are easily viewed.	_____
0 - 2	Labels	Sections of study design are labeled.	_____
0 - 2	Attractive	Uses color for emphasis, good arrangement, graphics	_____
0 - 2	Text on Display	Correct spelling and grammar, clear and concise writing	_____
0 - 2	Creative Approach	Evidence of researcher's original input into design	_____

EXPERIMENT STUDY

Points	Criterion	Explanation	Rating
0 -15	Log Book	A time-task recording of all steps of the study; recorded data	_____
0 -15	Background Information	History, significance, facts, and procedural information	_____
0 - 5	Problem	Question or statement of purpose about the relationship between independent variables (main characters) and dependent variables (plot)	_____
0 - 5	Hypothesis	Expected, directional relationship between independent variables (main characters) and dependent variables (plot)	_____
0 - 20	Procedure	Identifies the alterations of the selected independent variables (main characters), the dependent variables (plot) and how it will be measured, and the where, when, and how of the constant variables (setting); uses metric units	_____
0 -10	Trials and Samples	Appropriate number of samples and trials, use of controls	_____
0 -10	Results	Graph shows the relationship between independent variables and dependent variables and interpretation of the findings; uses averaged values	_____
0 - 5	Conclusions	Reaction to hypothesis consistent with results; includes link to background information and gives significance	_____
0 - 5	Scientific Worth	Thoroughness of plan, uses dry run, checks for valid and reliable data, gives possible future study	_____

FS-62105 Experiments